This journal belongs to

....

....

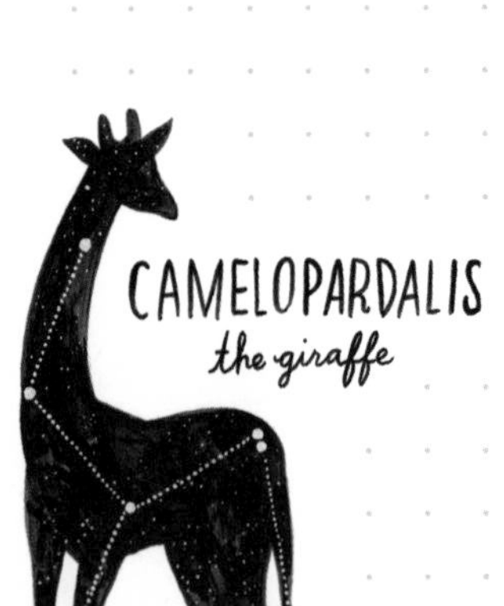
CAMELOPARDALIS
the giraffe

MONOCEROS
the unicorn

CANES VENATICI
the hunting dogs

APUS

the bird of paradise

CHAMELEON

the chameleon

N
S
N
S

Published in the United States by Clarkson Potter/Publishers,
an imprint of the Crown Publishing Group,
a division of Penguin Random House LLC, New York.
crownpublishing.com
clarksonpotter.com

Selected material originally appeared in *What We See in the Stars*
by Kelsey Oseid, published by Ten Speed Press,
an imprint of Penguin Random House LLC, New York, in 2017.

ISBN 978-1-9848-2317-5

Printed in China

Book design by Lise Sukhu
Illustrations by Kelsey Oseid Wojciak

First Edition